Thomas Andrews

Microscopic Internal Flaws Inducing Fracture in Steel

Thomas Andrews

Microscopic Internal Flaws Inducing Fracture in Steel

ISBN/EAN: 9783744686174

Printed in Europe, USA, Canada, Australia, Japan

Cover: Foto ©berggeist007 / pixelio.de

More available books at **www.hansebooks.com**

MICROSCOPIC INTERNAL FLAWS

INDUCING FRACTURE IN STEEL.

BY

THOMAS ANDREWS, F.R.S., F.C.S., M. Inst. C.E.

(TELFORD MEDALLIST AND TWICE PRIZEMAN,
INSTITUTION OF CIVIL ENGINEERS; ..., SOCIETY OF ENGINEERS),

CONSULTING METALLURGIST AND CHEMIST,

METALLURGICAL TESTING ... NEAR SHEFFIELD.

*From "*ENGINEERING*..."*

London:
E. AND F. N. SPON, ...

New York:
SPON AND CHAMBERLAIN, 12, CORTLANDT STREET.
1896.

MICROSCOPIC INTERNAL FLAWS

INDUCING FRACTURE IN STEEL.

BY

THOMAS ANDREWS, F.R.S., F.C.S., M. Inst. C.E.

(TELFORD MEDALLIST AND TELFORD PRIZEMAN,
INSTITUTION OF CIVIL ENGINEERS; BESSEMER PRIZEMAN, SOCIETY OF ENGINEERS),

CONSULTING METALLURGICAL ENGINEER AND CHEMIST,

METALLURGICAL TESTING LABORATORY, WORTLEY, NEAR SHEFFIELD.

REPRINTED

From "ENGINEERING," *July* 10, 17 *and* 24, 1896.

London:

E. AND F. N. SPON, 125, STRAND.

New York:

SPON AND CHAMBERLAIN, 12, CORTLANDT STREET.

1896.

MICROSCOPIC INTERNAL FLAWS INDUCING FRACTURE IN STEEL.

MICROSCOPIC INTERNAL FLAWS IN STEEL RAILWAY LOCOMOTIVE AND STRAIGHT AXLES, TYRES, RAILS, STEAMSHIP PROPELLER SHAFTS, AND PROPELLER CRANK SHAFTS, AND OTHER SHAFTS, BRIDGE GIRDER PLATES, SHIP PLATES, AND OTHER ENGINEERING CONSTRUCTIONS OF STEEL.

THE author has for many years past been engaged on researches with the object of ascertaining some of the causes of the deterioration of metals leading to the accidental fracture of railway axles, propeller shafts, and other metallic constructions. He has recently studied the "Effects of Temperature on the Strength of Railway Axles," in an investigation extending over ten years, and has determined, on a large experimental scale, the resistance of metals to sudden concussions at varying temperatures down to zero (0 deg.) Fahr., and indicated the influence of climatic temperature changes on the strength of railway material; the experiments showing that, under certain conditions, temperature is a potent factor in leading to the deterioration of the resisting strength of railway axles and shafts. The author has also completed another investigation on the "Effect of Strain on Railway Axles," in which some other aspects of the deterioration by fatigue in metals have been experimentally examined. If it were possible to produce a perfect metal, theoretically, there should be no deterioration by fatigue; but, alas! this at present seems almost impossible.

> Imperfections abound without, within,
> In toughest metals as metallic sin.

All, therefore, that the metallurgist and scientific investigator can do is:

1. To endeavour to find out the ultimate causes of internal defects in metals.

2. To minimise the deleterious influence of these, if possible, in existing structures.

3. On the manipulation of metals to endeavour in the

future to avoid by suitable and practical methods the possibility of internal flaws or defects arising.

Whatever be the cause of the deterioration of metals

FIG. 1.   Internal Micro-Flaw × 250 ; carbon 0.10 per cent., sulphur 2.0 per cent.

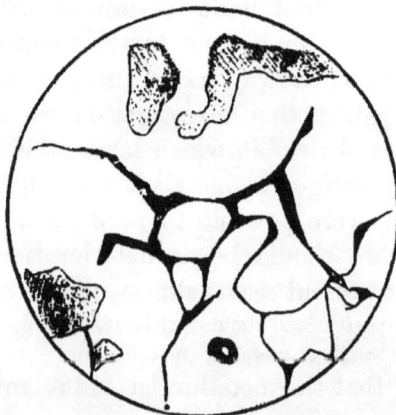

FIG. 2.   Internal Micro-Flaws × 300 ; sulphur 0.25 per cent.

through repetition of stress, or the fatigue of wear and tear (and notwithstanding the most valuable experiments of Wohler, Sir Benjamin B. Baker, Professor Bauschinger, Professor Kennedy, and others, comparatively little is known

of the true cause of such deterioration), there is a consensus
of facts indicating that a weakening of endurance occurs in

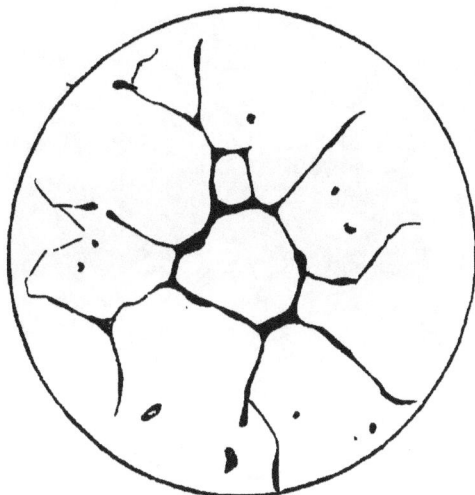

FIG. 3. Internal Sulphur Flaws × 300 diameter.

FIG. 4. Bessemer Steel Railway Axle, showing Internal Micro-Flaws × 500 diameter.

iron and steel through the action of repeated stress. In
connection with this interesting subject reference may be

made to some remarks by Professor W. C. Unwin, F.R.S., M. Inst. C.E.: "In all cases the number of repetitions of

FIG. 5. Internal Flaw × 300 diameter, Bessemer Steel Railway Axle 2.

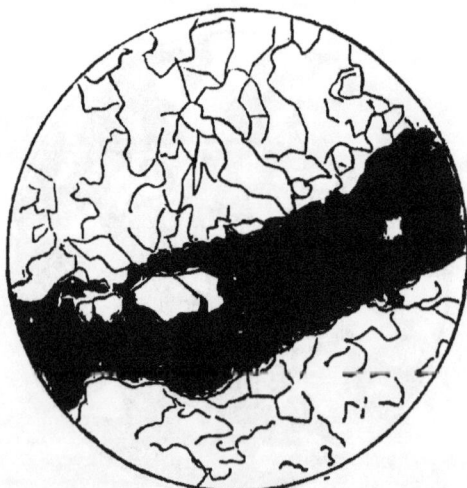

FIG. 6. Bessemer Steel Railway Axle, Internal Flaw × 300 diameter.

loading the bar will bear diminishes with increased range of variation of stress." "It is impossible not to conclude that, whatever the cause of decreased life of the bar may be, it is a cause which acts continuously, altering in some way the

Fig. 7. Internal Micro-Flaws, Belgian Bessemer Steel Railway Axle, × 800 diameter.

Fig. 8. Internal Micro-Flaws, Belgian Bessemer Steel Railway Axle, × 800 diameter.

structure or the properties of the bar." "The material, after a certain number of repetitions with a given range of stress,

does break with fewer subsequent repetitions.   For some
reason the ordinary testing machine observations are too

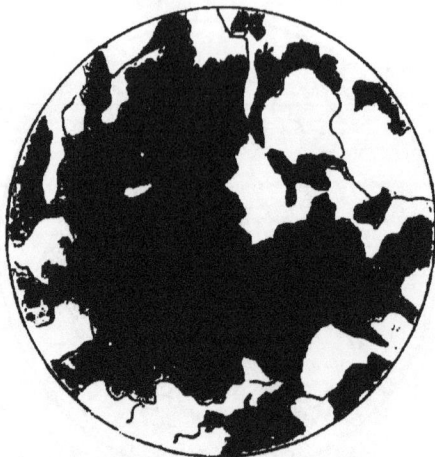

Fig. 9.   Siemens Steel (W.S.) Railway Axle, Internal Micro-Flaws, × 300 diameter.

Fig. 10.   Siemens Steel Railway Axle, showing Internal Micro-Flaws, × 500 diameter.

coarse to detect the difference."   "The deterioration may be
primarily a loss of power yielding in the particles near the
place of weakness, and not a loss of tenacity."   ("The Testing

of Materials of Construction," by Professor W. C. Unwin, F.R.S., M. Inst. C.E., pages 378 and 379.)

FIG. 11.   Siemens Steel Railway Axle, showing Internal Micro-Flaws, × 300 diameter.

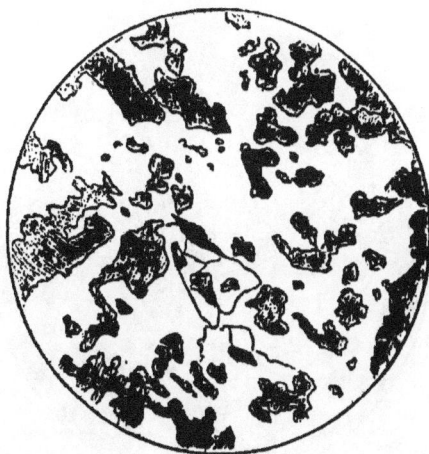

FIG. 12.   Siemens Steel Locomotive Crank Axle, showing Internal Micro-Flaws, × 300 diameter.

In the study of the causes leading to the deterioration by fatigue in metals, ultimately producing fracture, it seemed to the author desirable primarily to undertake a careful investigation with high microscopical powers, of the visible, tangible, and measurable causes influencing the enduring strength of

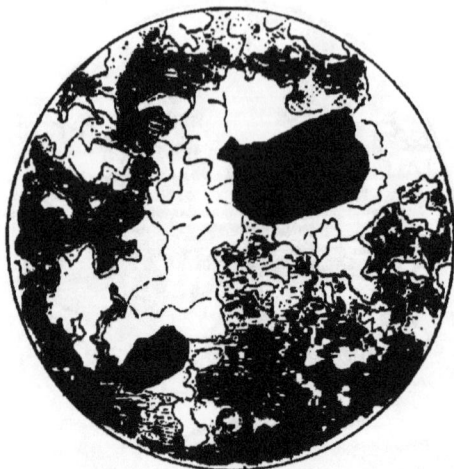

FIG. 13.   English Bessemer Steel Railway Tyre, Internal Micro-Flaws, × 400 diam.

FIG. 14.   Siemens Steel Railway Tyre, showing Internal Micro-Flaws, × 500 diam.

metals.   This method will prove of more practical value than any attempt at theorising in connection with such an important subject, and will doubtless lead to interesting and important results.

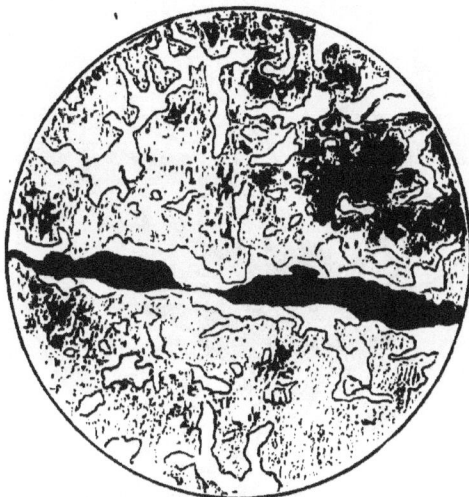

FIG. 15. Bessemer Steel Rail, 86 lb. per yard, showing Internal Micro-Flaws, × 300 diameter.

FIG. 16. Siemens Steel Propeller Shaft, Internal Micro-Flaw, × 500 diameter.

In view of the numerous serious accidents which have occurred in recent years from the sudden fracture of steel axles, rails, propeller shafts, &c., the author thought it would be further desirable to endeavour to investigate with high

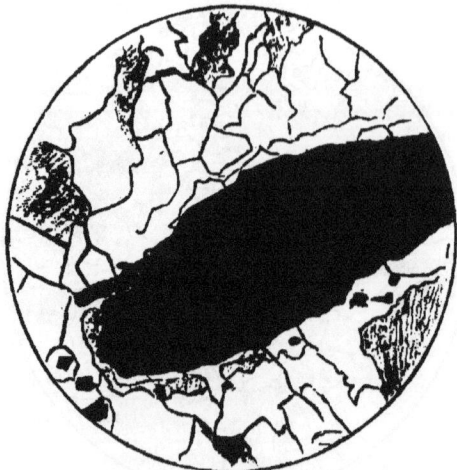

FIG. 17.   Internal Micro-Flaw, Siemens Steel Propeller Shaft, × 300 diameter.

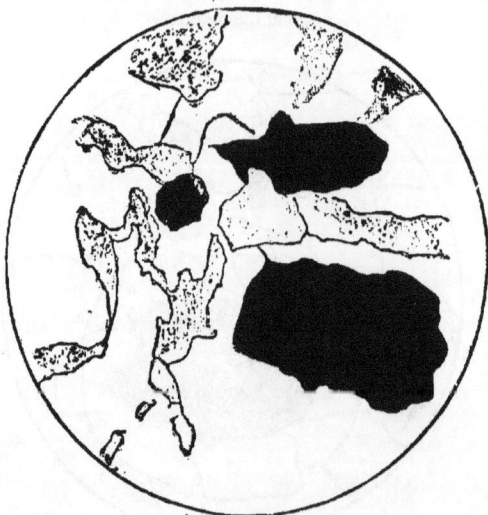

FIG. 18.   Internal Micro-Flaws, Siemens Steel Propeller Shaft, × 800 diameter.

microscopical powers the ultimate crystalline structure of steel axles and shafts, so as to obtain information, if possible, which might throw additional light on the source or sources of some of these so-called mysterious failures. With this object, the author has first directed his attention to a study of the structure of the internal micro-flaws which are almost invariably present in steel forgings, as these undoubtedly constitute a chief source of initial weakness in axles, rails and shafts, heavy guns, &c., often leading to their sudden fracture, or hastening the deterioration by fatigue. To carry out the research, the author employed a large compound binocular microscope, with specially arranged universal rack

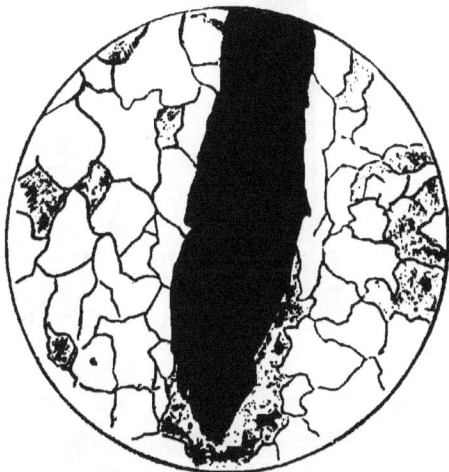

FIG. 19.   Internal Micro-Flaw, Siemens Steel (Warship) Propeller Shaft, × 300 diam.

motions so designed that the whole surface of a large fractured area could be traversed in any direction, and by this appara-tus he has been able to examine the fractured surface of large shafts or forgings microscopically with high powers, the microscope itself being arranged to travel over the surface of the fracture of any shaft or forging, however large, independ-ently of the size of the forging. By this means the author is able to locate and pursue any internal flaw from its source to its various ramifications, and he has obtained much valuable information from a study of large fractured surfaces by this means.

For more detailed examination, with still higher powers of the microscope, sections ($\frac{1}{2}$ in. in diameter, and sometimes larger) were machined out from railway axles, tyres, propeller shafts, crankshafts, heavy guns, or other large forgings, &c., and carefully prepared and etched in nitric acid of suitable strength for obtaining the most reliable development of the ultimate structure of the metals. The micro-sections were

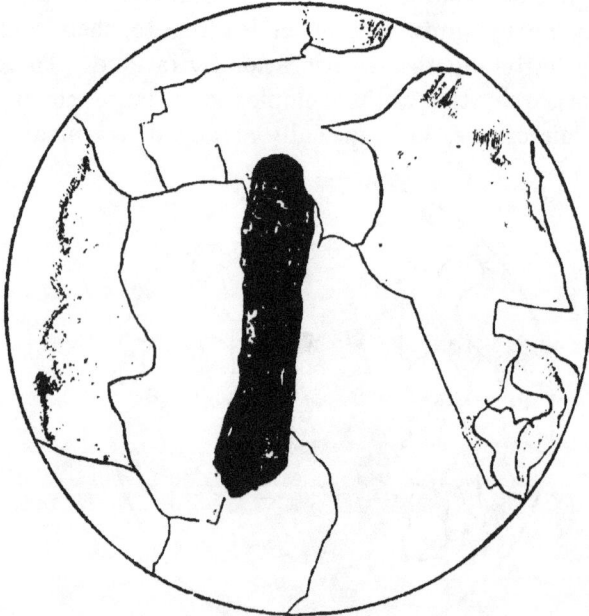

FIG. 20.   Internal Micro-Flaw, Siemens Steel Propeller Crankshaft, × 800 diam.

under constant observation with lenses during the etching process. Afterwards the metals were examined by another large compound microscope ·fitted with the author's special apparatus devised for this class of work.

The research is a difficult one, but the author has hope of obtaining information relating to some aspects of the subject which he trusts may prove useful.

The experiments were divided into Sets I., II., and III.

Set I.—Observations on the internal micro-flaws in Bessemer and Siemens steel railway axles, locomotive crank axles, tyres

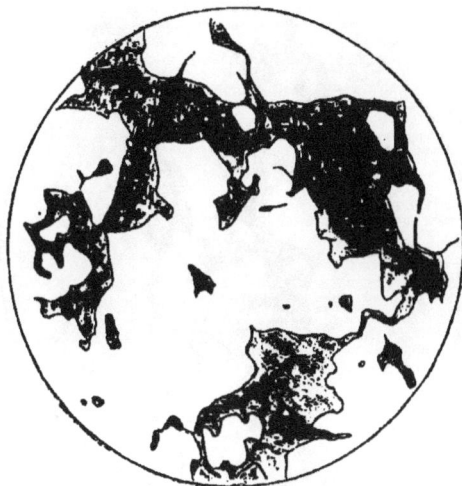

FIG. 21.   Internal Micro-Flaws, Siemens (soft) Steel Shaft, × 300 diameter.

FIG. 22.   Internal Micro-Flaws, Siemens (soft) Steel Shaft, × 300 diameter.

and rails, selected from the best makers in the various manufacturing districts in Great Britain and Belgium.

The axles and tyres were made in accordance with the requirements and tests of the British Railway companies.

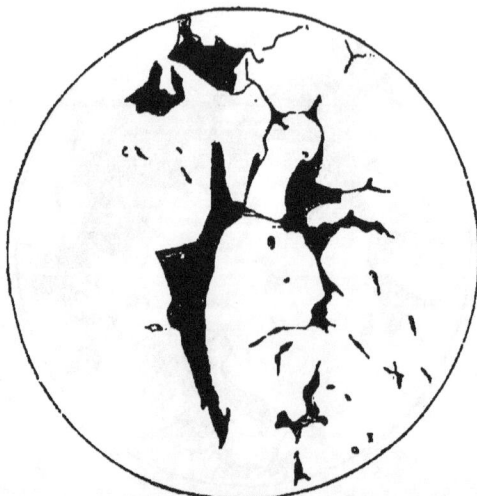

FIG. 23.   Internal Micro-Flaws, Siemens (soft) Steel Shaft, × 300 diameter.

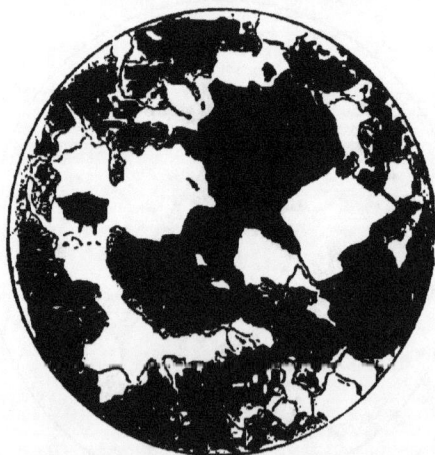

FIG. 24.   Siemens (hard) Steel Shaft, Internal Micro-Flaws, × 300 diameter.

Set II.—Observations on the internal micro-flaws in large Siemens steel steamship propeller shafts for war vessels and the mercantile marine; smaller steel shafts of soft Siemens steel, hard Siemens steel, soft Bessemer steel, aluminium steel, nickel steel, copper steel, silicon steel, and chromium steel, were also microscopically examined.

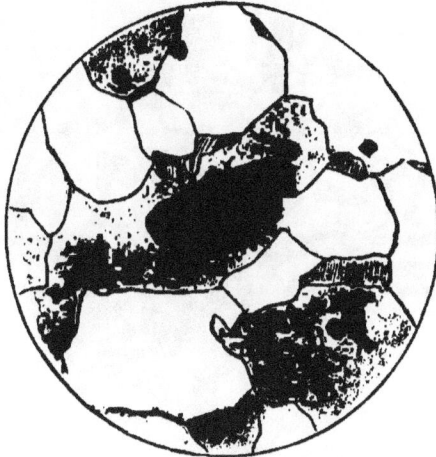

FIG. 25.   Internal Micro-Flaws, Bessemer (soft) Steel Shaft (rolled), × 500 diam.

FIG. 26.   Internal Micro-Flaws, Nickel Steel Shaft, × 500 diameter.

The propeller shafts examined (see Table VI.) varied in size from 36 ft. long by 12 in. diameter, weighing about six tons. Most of them were forged under heavy steam hammers, but in other instances the manipulation was done with a powerful hydraulic forging press. The propeller crankshaft (Table VII.) was about 12 in. diameter, with a stroke of 3 ft.

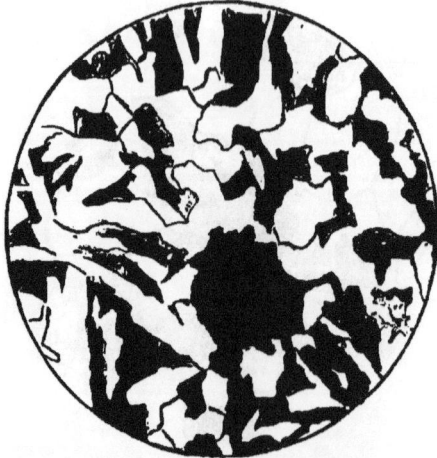

Fig. 27.   Copper Steel Shaft, Internal Micro-Flaws, × 300 diameter.

Fig. 28.   Silicon Steel Shaft, Internal Micro-Flaws, × 300 diameter.

Set III.—Observations on the internal micro-flaws in Siemens steel ship-plates, bridge girder plates, boiler plates, &c., varying in thickness from 1.125 in. downwards.

The author has not in this paper attempted to deal with extensive flaws, such as large blowholes, or gas cavities and pipe cavities in steel, though such are not infrequently present,

Fig. 29.   Internal Micro-Flaws, Siemens Steel Ship-plate, × 300 diameter.

Fig. 30.   Siemens Steel Boiler Plate, showing Internal Micro-Flaws, × 300 diameter.

but he has limited his observations to the more numerous subtle and minute sources of internal weakness in constructive metals. The internal micro-flaws in the steel axles, propeller shafts, &c., observed by the author, varied in size from about 0.0360 in. long, 0.0080 in. wide, as seen in section. (See Tables.) It was, however, impossible to ascertain the distance they extended into the forgings, or the full extent of their various ramifications. Such portions only of the micro-flaws as were cut by the section could be accurately observed.

The dimensions of the micro-flaws were carefully taken by a Jackson micrometer, and in some cases with a Ramsden screw micrometer, both accurately calibrated with a standard stage micrometer.

In addition to blowholes, air cavities, and other causes of weakness, the microscope has recently revealed a further source of internal and growing flaws in steel axles and shafts, heavy guns, &c. A careful high-power microscopical exami-nation of steel in which sulphur is present, shows that the sulphur is combined with the iron, forming a sulphide of iron.* During the cooling and crystallisation of the metal the minute quantity of sulphide of iron present has been found, by the use of high microscopic powers, to have chiefly located itself in the intercrystalline spaces forming the ultimate structure of the metal. In steel castings, the sulphide of iron is mostly found in long thin veins in the intercrystalline spaces. In large forgings, or other steel structures which have either been reheated or annealed at suitable tempera-tures, the sulphide of iron often segregates, drawing itself together, as it were, into ovoid nodules. The intercrystalline spaces thus left vacant, either metallically weld together, if the temperature of subsequent heating or annealing is sufficient, or they remain as fine fissures, forming sources of internal weakness in the metal. (Professor Arnold has made a special study of the influence of sulphur and other impurities on steel castings, and the author's own experimental results are in accord with, and confirmatory of, Professor Arnold's observations.) This becomes a constant source of weakness,

* When manganese is present in excess, a sulphide of manganese is also formed. The micro-sulphide flaws are generally of the ovoid dove-coloured type when manganese is present in excess,

increasing or diminishing in proportion to the percentage of sulphur present in the metal. It is, however, rarely wholly absent. Such interference with the natural cohesion between the ultimate crystalline facets of the metal has a tendency to induce serious lines of internal weakness in the structure, and constitutes an element of danger. These internal interferences with the natural intercrystalline spaces are frequently very minute, and in fact are undetectable by the eye; but when specially prepared and etched sections of the steel are examined under high microscopical powers, these minute sources of initial weakness are observable. (See micro-measurements on the Tables, and the illustrations in Figs. 1 to 30, pages 4 to 19.)

Dr. H. C. Sorby, F.R.S., was the first to commence the microscopical study of the structure of iron and steel, and Professor J. O. Arnold, F.C.S., of the Sheffield Technical School, was, the author believes, the first to microscopically study the deleterious effect of sulphur and other impurities on steel, in his recent most valuable contribution to the metallurgy of steel, entitled, "The Physical Influence of Elements on Iron," read at the meeting of the Iron and Steel Institute, in May, 1894.

Valuable metallurgical microscopic researches have also been made by Dr. Wedding, Chernoff, Bayles, MM. Osmond and Werth, Dr. Martens, A. Sauveur, and others.

In the course of his paper Professor Arnold pertinently remarks that: "The large crystals appeared to be of one type, through which the sulphides of iron seemed fairly evenly distributed, but they evidently possess an extraordinary power of individual contraction, the result being that large fissures are developed between the joints in such numbers that the metal is almost cut into pieces. In some instances actually detached crystals may be observed, and many such fell out during the compression test in the form of a silvery dust. It is extremely improbable that the intercrystalline spaces in cold steel, when due to the presence of sulphur, will weld up on heating; hence the red-shortedness also produced

by the presence of considerable quantities of this element.
This micro-section forms a good example of the many cases
occurring in metallurgy in which the study of molecules
alone is of secondary importance when compared with the
investigation of the junction lines of collective masses containing innumerable molecules and grouped into crystals."

The author of the present paper has also made numerous
microscopical observations, having examined various large
forgings, such as Bessemer steel railway axles, Siemens steel
axles. and tyres, manufactured by the leading makers in
various parts of Great Britain. Some steel axles and tyres
of Belgian make which came under the author's cognisance
were also examined, and also large Siemens steel propeller
shafts of British manufacture for war vessels and the mercantile marine, varying in weight from about 6 tons. The
author has further made microscopical examinations of large
artillery guns (37 tons in weight), projectiles, ship plates,
bridge plates, boiler plates, &c., and other constructive
materials which have come under his observation. His own
microscopic researches on the above-mentioned constructive
metals fully coincide with Professor Arnold's remarks and
experiments which were made on smaller masses of steel.

The physical tests, chemical composition, &c., of the metal
experimented upon are given on Tables I. and II.

The whole of the steel axles and tyres were made in
accordance with the requirements and satisfied the tests of
the British railway companies for this class of material, and
the steel propeller shafts, bridge plates, &c., satisfied the tests
imposed.

The author's microscopic observations were taken with a
$\frac{1}{8}$ in., $\frac{1}{10}$ in., and other objectives, the magnifying powers
varying from 300 diameters, and extending upwards to 2,000
diameters. Careful micrographs and drawings were also
made by the author of the structure of the various internal
micro-flaws under notice.

It may, perhaps, be desirable briefly to explain the
illustrations given in Figs. 1 to 30, pages 4 to 19. As

TABLE I.—CHEMICAL ANALYSES OF THE STEEL RAILWAY AXLES, TYRES, RAILS, STEAMSHIP PROPELLER SHAFTS, PROPELLER CRANKSHAFTS AND OTHER SHAFTS, SHIP AND BOILER PLATES, BRIDGE GIRDER PLATES, &C., EXAMINED IN THE EXPERIMENTS.

Percentage Results.

	Bessemer Steel Railway Axle, British Make. Index No. 8 A.	Siemens Steel Railway Axle, British Make. Index No. 6 A.	Siemens Steel Railway Axle, British Make. Index No. 8 A.	Bessemer Steel Railway Tyre, British Make. Index No. 11 A.	Siemens Steel Railway Tyre, British Make. Index No. 12 A.	Bessemer Steel Railway Tyre, Belgian Make. Index No. 13 A.	Siemens Steel Propeller Shaft. Index No. 15 A.	Siemens Steel Propeller Shaft. Index No. 16 A.	Siemens Steel Propeller Crankshaft. Index No. 20 A.	Soft Siemens Steel Shaft. Index No. 21 A.	Hard Siemens Steel Shaft. Index No. 22 A.	Soft Bessemer Steel Shaft. Index No. 23 A.
Combined carbon	0.410	0.400	0.380	0.340	0.350	0.420	0.170	0.290	0.290	0.230	0.460	0.150
Silicon	0.062	0.118	0.080	0.069	0.067	0.102	0.020	0.160	0.047	0.014	0.107	0.009
Manganese	0.740	0.898	0.763	1.009	0.960	0.694	1.170	0.530	0.895	0.698	0.972	0.468
Sulphur	0.080	0.110	0.060	0.130	0.050	0.070	0.040	0.027	0.070	0.100	0.023	0.112
Phosphorus	0.064	0.074	0.058	0.086	0.044	0.070	0.052	0.034	0.064	0.075	0.075	0.088
Iron (by difference)	98.644	98.472	98.659	98.366	98.519	98.644	98.548	98.969	98.674	98.883	98.363	99.173
	100.000	100.000	100.000	100.000	100.000	100.000	100.000	100.000	100.000	100.000	100.000	100.000

TABLE I.—*continued.*

	Aluminium Steel Shaft. Index No. 24 A.	Nickel Steel Shaft. Index No. 25 A.	Copper Steel Shaft. Index No. 26 A.	Silicon Steel Shaft. Index No. 27 A.	Chromium Steel Shaft. Index No. 28 A.	Siemens Steel Ship Plate. Index No. 29 A.	Siemens Steel Bridge Girder Plate. Index No. 30 A.	Siemens Steel Boiler Plate. Index No. 32 A.	Siemens Steel Locomotive Crank Axle, British Make. Index No. 10 A.	Siemens Steel Warship Propeller Shaft. Index No. 18 A.	Siemens Steel Propeller Shaft. Index No. 17 A.	Bessemer Steel Ball Section, 86 lb. per Yard. Index No. 14 A.
Combined carbon	0.240	0.210	0.220	0.240	0.800	0.120	0.150	0.130	0.350	0.430	0.310	0.440
Silicon	0.560	0.580	0.610	0.570	:	0.027	0.021	0.026	0.093	0.053	0.070	0.040
Manganese	0.820	1.000	0.930	0.900	:	0.178	0.521	0.680	0.661	1.146	0.620	0.800
Sulphur	0.070	0.070	0.070	0.070	:	0.047	0.045	0.046	0.030	0.050	0.022	0.100
Phosphorus	0.070	0.070	0.070	0.070	:	0.068	0.064	0.078	0.040	0.064	0.036	0.064
Aluminium	0.420											
Nickel		1.960										
Copper			0.490									
Chromium					2.210							
Iron (by difference)	97.820	96.110	97.610	98.150	96.290	99.560	99.199	99.091	98.926	98.252	98.942	98.556
	100.000	100.000	100.000	100.000	100.000	100.000	100.000	100.000	100.000	100.000	100.000	100.000

previously mentioned, these are accurate micro-drawings. The black and heavily shaded (for convenience) portions of each drawing are the internal micro-flaws, the lighter shaded parts are the ultimate primary crystals composed of distinct and detached crystalline areas of iron, saturated to various extents with combined carbon. These may appositely be called normal carbide of iron areas or crystals, as distinct from the pure iron crystals (or ferrite) of the metal. The unshaded portions of the drawings consist of crystals of pure iron, or ferrite, which have not been affected, or but very slightly, by the carbon. It may be observed that these areas of pure iron crystals are found to diminish in quantity in steels as the percentage of combined carbon in the steel

TABLE II.—PHYSICAL PROPERTIES OF THE STEEL RAILWAY AXLES, TYRES, STEAMSHIP PROPELLER SHAFTS, PROPELLER CRANKSHAFTS AND OTHER SHAFTS, SHIP AND BOILER PLATES, BRIDGE GIRDER PLATES, &C., EXAMINED IN THE EXPERIMENTS.

DESCRIPTION.	Index Number.	Original Dimensions of Test Piece, Diameter in Inches.	Distance between Datum Points in Inches.	Breaking Stress in Tons per Square Inch.	Elongation per Cent.	Reduction of Area per Cent.
Bessemer steel railway axle, British make	2 A	.798	3	36 42	30.0	
Ditto ditto	3 A	.798	3	37.86	26.7	47.2
Bessemer steel railway axle, Belgian make	4 A	.798	3	40.70	25 0	
Siemens steel railway axle, British make	6 A	.798	3	43.70	20.0	31.6
Ditto ditto	7 A	.798	3	24.60	27.0	33.6
Ditto ditto	8 A	.798	3	35 00	29.0	
Ditto ditto	9 A	.785	3	37.80	31.6	
Siemens steel locomotive crank axle, British make	10 A	.798	3	26.92	34.7	57.6
Bessemer steel railway tyre, British make	11 A	.799	3	38.21	28.0	
Siemens steel railway tyre, British make	12 A	.798	3	37·00	26.0	
Bessemer steel railway tyre, Belgian make	13 A	.798	3	39.00	25.7	
Siemens steel propeller shaft	15 A	.798	3	28.44	25.0	41.6
Ditto	16 A	.798	3	30.26	27.3	45 4
Ditto	17 A	.798	3	31 42	36 0	60.8
Siemens steel warship propeller shaft	18 A	.798	3	30.76	32.3	57.6
„ propeller crankshaft	20 A	.793	8	27.64	30.7	54.2
Soft Siemens steel shaft	21 A	.559	3	28.90	28.0	35.1
Hard Siemens steel shaft	22 A	.875	3	50.07	14.0	48.4
Soft Bessemer steel shaft	23 A	.559	3	26 29	40.0	53.9
Aluminium steel shaft	24 A	.559	3	33.39	32.0	40.8
Nickel steel shaft	25 A	.559	3	37.59	38.0	53.8
Copper steel shaft	26 A	.875	3	35.72	20.0	26.5
Silicon steel shaft	27 A	.559	3	34.12	37.0	48.6
Chromium steel shaft	28 A	.875	3	45.69	none	none
Siemens steel ship-plate	29 A	.564	3	25.71	31.6	79.2
„ bridge girder plate	30 A	.600	3	27.12	31.6	61.7
„ boiler plate	32 A	.395	3	30.00	19.2	59.8

DESCRIPTION OF THE STEEL RAILWAY LOCOMOTIVE AND STRAIGHT AXLES, TYRES, RAILS, STEAMSHIP PROPELLER SHAFTS AND PROPELLER CRANKSHAFTS, AND OTHER SHAFTS, BRIDGE GIRDER PLATES, SHIP-PLATES OR OTHER STEEL ENGINEERING CONSTRUCTIONS EXAMINED IN THIS RESEARCH.

Reference to Illustrations.	Reference to Tables.	Index Number	DESCRIPTION.
Fig. 4, page 5	Set I., Table III.	2 A	Bessemer steel railway axle. British manufacture
,, 5, ,, 6	,, I., ,, III.	3 A	Ditto. Ditto.
,, 7, ,, 7	,, I., ,, III.	4 A	Ditto. Belgian manufacture.
,, 8, ,, 7	,, I., ,, III.	5 A	Ditto. Ditto.
	,, I., ,, IV.	6 A	Siemens steel railway axle. British manufacture.
,, 9, ,, 8	,, I., ,, IV.	7 A	Ditto. Ditto.
,, 10, ,, 8	,, I., ,, IV	8 A	Ditto. Ditto.
,, 11, ,, 9	,, I., ,, IV.	9 A	Ditto. Ditto.
,, 12, ,, 9	,, I., ,, IV.	10 A	Siemens steel locomotive crank axle. British manufacture.
,, 13, ,, 10	,, I., ,, V.	11 A	Bessemer steel railway tyre. British manufacture
,, 14, ,, 10	,, I., ,, V.	12 A	Siemens ,, ,, Ditto.
	,, I., ,, V	13 A	Bessemer ,, ,, Belgian manufacture.
,, 15, ,, 11	,, I., ,, V.	14 A	,, steel rail. British manufacture.
,, 16, ,, 11	,, II., ,, VI.	15 A	Siemens steel steamship propeller shaft of British manufacture, size 30 ft. long and about 12 in. in diameter, forged under a steam hammer.
	,, II., ,, VI	16 A	Siemens steel steamship propeller shaft, British manufacture, size 36 ft. long and 12 in. diameter, made by a very powerful hydraulic forging press.
	,, II., ,, VI.	17 A	Siemens steel steamship propeller shaft, British manufacture, forged under large steam hammer
,, 17, ,, 12	,, II., ,, VI.	18 A	Siemens steel warship propeller shaft of British manufacture, forged under a steam hammer.
,, 18, ,, 12	,, II., ,, VI.	19 A	Siemens steel warship propeller shaft of British manufacture, forged under a steam hammer.
,, 20, ,, 14	,, II., ,, VII.	20 A	Siemens steel steamship propeller crankshaft, British manufacture. Diameter of shaft about 12 in., crank 18 in., with a stroke of 3 ft., forged under a steam hammer.
,, 21, ,, 15	,, II., ,, VII.	21 A	Soft Siemens steel rolled shaft, 3 in. diameter.
,, 24, ,, 16	,, II., ,, VII.	22 A	Hard Siemens ,, 3 ,,
,, 25, ,, 17	,, II., ,, VII.	23 A	Soft Bessemer ,, 3 ,,
	,, II., ,, VIII.	24 A	Aluminium steel rolled shaft, $3\frac{1}{4}$ in. in diameter by 5 ft. long.
,, 26, ,, 17	,, II., ,, VIII.	25 A	Nickel steel rolled shaft, $3\frac{1}{4}$ in. diameter, 5 ft. long.
,, 27, ,, 18	,, II., ,, VIII.	26 A	Copper ,, $3\frac{1}{4}$,, 5 ,,
,, 28, ,, 18	,, II., ,, VIII.	27 A	Silicon ,, $3\frac{1}{4}$,, 5 ,,
	,, II., ,, VIII.	28 A	Chromium ,, $3\frac{1}{4}$,, 5 ,,
,, 29, ,, 19	,, III., ,, IX.	29 A	Siemens steel ship-plate, $1\frac{1}{8}$ in. thick.
	,, III., ,, IX.	30 A	,, bridge girder plate, $\frac{3}{8}$ in. thick.
,, 30, ,, 19	,, III., ,, IX.	31 A	,, boiler plate, $\frac{1}{2}$ in. thick.
	,, III., ,, IX.	32 A	,, ,, $\frac{1}{2}$,,

It may be remarked that the various steel axles, propeller shafts, and other engineering constructions of steel examined were of the most modern type, and were made by manufacturers of the highest repute in various parts of Great Britain and on the Continent. Many of them were made in the year 1894.

Illustrations of typical internal micro-flaws solely due to the presence of excess of sulphur in iron or steel will be seen on reference to the micro-sections, Figs. 1, 2, and 3, pp. 4, 5. These flaws were due entirely to the presence of excess of sulphur purposely alloyed in samples of pure iron made by Professor J. O. Arnold, F.C.S., Sheffield Technical School. The analysis of the metal in which these sulphur flaws were found was as follows: The sample from which micro-section Fig. 1, page 4, was taken contained iron 97.90 per cent.; combined carbon, 0.10 per cent.; sulphur, 2.00 per cent. The samples from which the micro-sections Figs. 2 and 3, pp. 4, 5, were taken contained iron about 99.65 per cent., combined carbon, 0.10 per cent.; sulphur, 0.25 per cent. The incipient micro-flaws due to sulphur in these samples varied in size from about 0.018 in. in length, and in thickness from about 0.001 in., tapering down to as small as 0.00001 in.

The exact dimensions of the sulphur micro-flaws taken from micro-sections of the last two samples mentioned above, viz., those containing 0.25 per cent. of sulphur, are given on Set I., Table III. These flaws may be regarded as typical of those produced by the presence of excess of sulphur in iron and steel, though the form assumed by these sulphur flaws varies somewhat in different specimens of steel.

increases until the saturation point of 0.89 per cent. of carbon is reached, when the iron crystals, which have up to this point remained unacted upon by the carbon, disappear, and the whole area becomes filled with grey or darker crystals of iron carbide (these carbide of iron areas often containing the

SET I.—TABLE III.—Measurements of Internal Micro-Flaws in Bessemer Steel Railway Axles, &c.

Typical Internal Micro-Flaws in Iron or Steel owing to the Presence of Sulphur. Index No. 1 A. (See Fig. 1, page 4.)			Bessemer Steel Railway Axle, British Make. Index No. 2 A. (See Fig. 4, p. 5.)		Bessemer Steel Railway Axle, British Make. Index No. 3 A*. (See Fig. 5, page 6.)		Bessemer Steel Railway Axle, Belgian Make. Index No. 4 A. (See Fig. 7, page 7.)		Bessemer Steel Railway Axle, Belgian Make. Index No. 5 A. (See Fig. 8, page 7.)	
Maximum Longitudinal Dimensions of Internal Micro-Flaws.	Maximum Width of Internal Micro-Flaws.	Minimum Width of Internal Micro-Flaws.	Longitudinal Dimensions of Internal Micro-Flaws.	Transverse Dimensions of Internal Micro-Flaws.	Longitudinal Dimensions of Internal Micro-Flaws.	Transverse Dimensions of Internal Micro-Flaws.	Longitudinal Dimensions of Internal Micro-Flaws.	Transverse Dimensions of Internal Micro-Flaws.	Longitudinal Dimensions of Internal Micro-Flaws.	Transverse Dimensions of Internal Micro-Flaws.
					Dimensions in Parts of an Inch.					
.0060	.0004	.000020	.0012	.0010	.0300	.0040	.0012	.0002	.0006	.0004
.0050	.0004	.000020	.0014	.0006	.0008	.0004	.0004	.0003	.0008	.0004
.0088	.0002	.000020	.0012	.0004	.0002	.0002	.0006	.0003	.0004	.0004
.0030	.0002	.000013	.0004	.0002	.0004	.0004	.0004	.0002	.0006*	.0004
.0050	.0002	.000013	.0006	.0004	.0004	.0002	.0006	.0004	.0006	.0004
:0110	.0002	.000017	.0006	.0004	.0004	.0002	.0006	.0002	.0008	.0004
.0084	.0002	.000010	.0003	.0004	.0006	.0004	.0006	.0003	.0006	.0004
.0140	.0002	.000019	.0004	.0002	.0004	.0004	.0004	.0002	.0006	.0004
.0048	.0006	.000017	.0010	.0008	.0008	.0004	.0010	.0006	.0008	.0004
.0034	.0006	.000013	.0004	.0004	.0004	.0004	.0006	.0004	.0010	.0008
.0050	.0004	.000013	.0010	.0004	.0006	.0004	.0004	.0004	.0008	.0008
.0100	.0006	.000019	.0004	.0003	.0004	.0104	.0006	.0003	.0006	.0004
.0030	.0004	.000046	.0006	.0004	.0006	.0004	.0006	.0003	.0010	.0004
.0014	.0008	.000056	.0010	.0006	.0034	.0028	.0008	.0004	.0006	.0004
.0024	.0006	.000026	.0012	.0006	.0044	.0010	.0004	.0004	.0008	.0004
.0018	.0010	.000122	.0014	.0004	.0028	.0008	.0010	.0005	.0008	.0004
.0080	.0008	.000030	.0006	.0002	.0028	.0004	.0004	.0002	.0014	.0006
.0120	.0006	.000024	.0006	.0003	.0010	.0002	.0008	.0002	.0008	.0004
.0180	.0008	.000019	.0024	.0010	.0028	.0012	.0004	.0003	.0036	.0014
.0100	.0004	.0C0020	.0012	.0006			.0006	.0004	.0034	.0004
Highest .0180	.0010	..	.0024	.0010	.0300	.0040	.0012	.0006	.0036	.0014

* A micro-flaw of considerable extent was observed in this Bessemer steel railway axle, it being 0.03 in. long and 0.004 in. wide.

Fe$_3$C crystallised in fine parallel plates, alternating with fine plates of pure iron, not more than $\frac{1}{40000}$ in. to $\frac{1}{80000}$ in. apart). The fine divisional lines in the drawings show the line of junction or intercrystalline spaces between the ultimate crystals of the metals. With these few explanatory remarks the illustrations will be easily understood.

On reference to these illustrations, it will be seen in some cases that many of the ultimate crystals of the metal are separated and entirely isolated from each other by masses of sulphide of iron (sometimes of a dark shade, but more frequently having a dove-coloured appearance) which have

SET I.—*Continued.*—TABLE IV.—MEASUREMENTS OF INTERNAL MICRO-FLAWS IN SIEMENS STEEL RAILWAY AXLES AND LOCOMOTIVE CRANK AXLES.

Siemens Steel Railway Axle, British Make. Index No. 6 A.		Siemens Steel Railway Axle, British Make. Index No. 7 A. (See Fig. 9, page 8.)		Siemens Steel Railway Axle, British Make. Index No. 8 A. (See Fig. 10, page 8.)		Siemens Steel Railway Axle, British Make. Index No. 9 A. (See Fig. 11, page 9.)		Siemens Steel Locomotive Crank Axle, British Make. Index, No. 10 A. (See Fig. 12, page 9.)	
Longitudinal Dimensions of Internal Micro-Flaws.	Transverse Dimensions of Internal Micro-Flaws.	Longitudinal Dimensions of Internal Micro-Flaws.	Transverse Dimensions of Internal Micro-Flaws.	Longitudinal Dimensions of Internal Micro-Flaws.	Transverse Dimensions of Internal Micro-Flaws.	Longitudinal Dimensions of Internal Micro-Flaws.	Transverse Dimensions of Internal Micro-Flaws.	Longitudinal Dimensions of Internal Micro-Flaws.	Transverse Dimensions of Internal Micro-Flaws.
Dimensions in Parts of an Inch.									
.0012	.0006	.0032	.0010	.0006	.0003	.0100	.0010	.0006	.0004
.0006	.0004	.0028	.0010	.0004	.0004	.0024	.0008	.0008	.0004
.0010	.0004	.0050	.0008	.0006	.0004	.0060	.0004	.0008	.0004
.0004	.0004	.0054	.0010	.0006	.0004	.0040	.0002	.0004	.0002
.0008	.0001	.0012	.0010	.0006	.0004	.0018	.0002	.0008	.0006
.0008	.0002	.0018	.0016	.0004	.0003	.0098	.0006	.0008	.0004
.0003	.0001	.0100	.0060	.0008	.0004	.0020	.0002	.0006	.0002
.0006	.0004	.0006	.0004	.0008	.0004	.0078	.0002	.0012	.0003
.0006	.0002	.0008	.0008	.0004	.0004	.0028	.0002	.0010	.0004
.0008	.0004	.0024	.0008	.0014	.0006	.0120	.0002	.0010	.0004
.0006	.0004	.0060	.0006	.0008	.0004	.0060	.0004	.0006	.0002
.0006	.0003	.0030	.0004	.0007	.0004	.0018	.0004	.0006	.0004
.0006	.0004	.0020	.0004	.0006	.0004	.0026	.0004	.0004	.0004
.0008	.0002	.0018	.0004	.0008	.0004	.0050	.0002	.0010	.0004
.0006	.0005	.0024	.0004	.0006	.0004	.0070	.0002	.0008	.0004
.0005	.0002	.0032	.0010	.0008	.0004	.0040	.0002	.0006	.0002
.0007	.0003	.0060	.0010	.0004	.0004	.0062	.0002	.0008	.0002
.0007	.0002	.0034	.0008	.0006	.0004	.0046	.0006	.0004	.0002
.0009	.0002	.0024	.0006	.0028	.0014	.0048	.0004	.0008	.0004
.0010	.0016	.0052	.0010	.0010	.0010	.0100	.0002	.0008	.0004
0014	.0006	.0058	.0012	.0050	.0010				
Highest .0014	.0006	.0100	.0060	.0060	.0014	.0120	.0010	.0012	0.006

formed initial flaws and extensive lines of internal weakness in the metal. Other of the micro-flaws are hollow cavities, or minute blowholes, which have become distorted in course of forging and manipulation, and which sometimes contain slag. In many instances the normal intercrystalline spaces, and junctions, have also been disturbed and distorted owing to the presence of these flaws.

The sizes of the various micro-flaws or sources of incipient fracture in the metal are given respectively on Tables III. to IX.

The deleterious effect of these treacherous sulphur areas and other microscopic flaws, with their prolonged ramifications spreading along the intercrystalline spaces of the ultimate

SET I.—*Continued.*—TABLE V.—MEASUREMENTS OF INTERNAL MICRO-FLAWS IN BESSEMER AND SIEMENS STEEL RAILWAY TYRES AND RAILS.

Bessemer Steel Railway Tyre, British Make. Index No. 11 A. (See Fig. 13, page 10.)		Siemens Steel Railway Tyre, British Make. Index No. 12 A. (See Fig. 14, page 10.)		Bessemer Steel Railway Tyre, Belgian Make. Index No. 13 A.		Bessemer Steel Rail, British Make. Section 86 lb. per Yard. Index No. 14 A. (See Fig. 15, page 11.)	
Longitudinal Dimensions of Internal Micro-Flaws.	Transverse Dimensions of Internal Micro-Flaws.	Longitudinal Dimensions of Internal Micro-Flaws.	Transverse Dimensions of Internal Micro-Flaws.	Longitudinal Dimensions of Internal Micro-Flaws.	Transverse Dimensions of Internal Micro-Flaws.	Longitudinal Dimensions of Internal Micro-Flaws.	Transverse Dimensions of Internal Micro-Flaws.
Dimensions in Parts of an Inch.							
.0008	.0004	.0008	.0005	.0020	.0003	.0150	.0012
.0006	.0004	.0006	.0004	.0008	.0004	.0120	.0010
.0006	.0004	.0006	.0004	.0006	.0002	.0012	.0008
.0008	.0002	.0008	.0002	.0006	.0002	.0012	.0004
.0008	.0002	.0008	.0006	.0004	.0001	.0010	.0006
.0006	.0004	.0006	.0004	.0006	.0003	.0010	.0006
.0006	.0002	.0006	.0004	.0004	.0002	.0008	.0004
.0008	.0002	.0008	.0008	.0004	.0002	.0012	.0004
.0008	.0004	.0008	.0004	.0004	.0002	.0010	.0004
.0008	.0004	.0006	.0003	.0004	.0002	.0030	.0002
.0006	.0004	.0010	.0004	.0004	.0002	.0012	.0006
.0008	.0002	.0008	.0004	.0006	.0002	.0016	.0008
.0008	.0002	.0008	.0006	.0004	.0002	.0016	.0004
.0006	.0004	.u008	.0004	.0008	.0002	.0008	.0004
.0010	.0004	.0008	.0004	.0004	.0002	.0016	.0006
.0006	.0006	.0008	.0002	.0006	.0002	.0012	.0004
.0008	.0005	.0012	.0002	.0004	.0002	.0010	.0006
.0008	.0006	.0004	.0004	.0004	.0002	.0020	.0006
.0006	.0004	.0008	.0004	.0008	.0006	.0010	.0004
.0008	.0004	.0009	.0004	.0008	.0004	.0016	.0004
		.0008	.0006				
Highest .0010	.0006	0012	.0008	.0020	.0006	0 150	.0012

crystals of the metal, and destroying metallic cohesion, will be easily understood.

Constant vibration gradually loosens the metallic adherence of the crystals, especially in areas where these micro-flaws exist. Cankering by internal corrosion and disintegration is induced whenever the terminations of any of the sulphide areas or other flaws in any way become exposed at the surface of the metal either to the action of sea water, or atmospheric or other oxidising influences. In many other ways also it will be seen how deleterious is their presence. In the Tables (see pages 23 to 33) the author has recorded the results of observations on about twenty internal micro-flaws in each case, taken within a sectional area of only

SET II.—TABLE VI.—MEASUREMENTS OF INTERNAL MICRO-FLAWS IN LARGE SIEMENS STEEL STEAMSHIP PROPELLER SHAFTS FOR WAR VESSELS AND THE MERCANTILE MARINE.

Siemens Steel Propeller Shaft. Index No. 15 A. (See Fig. 16, page 11.)		Siemens Steel Propeller Shaft. Index No. 16 A.		Siemens Steel Propeller Shaft. Index No. 17 A.		Siemens Steel WarshipPropeller Shaft. Index No. 18A. (See Fig. 17, page 12.)		Siemens Steel Warship Propeller Shaft. Index No.19 A. (See Fig. 18, page 12.)	
Longitudinal Dimensions of Internal Micro-Flaws.	Transverse Dimensions of Internal Micro-Flaws.	Longitudinal Dimensions of Internal Micro-Flaws.	Transverse Dimensions of Internal Micro-Flaws.	Longitudinal Dimensions of Internal Micro-Flaws.	Transverse Dimensions of Internal Micro-Flaws.	Longitudinal Dimensions of Internal Micro-Flaws.	Transverse Dimensions of Internal Micro-Flaws.	Longitudinal Dimensions of Internal Micro-Flaws.	Transverse Dimensions of Internal Micro-Flaws.
Dimensions in Parts of an Inch.									
.0030	.0014	.0018	.0008	.0004	.0004	.0038	.0016	.00366	.00166
.0020	.0014	.0012	.0004	.0004	.0004	.0032	.0010	.00200	.00083
.0010	.0010	.0040	.0004	.0016	.0014	.0014	.0010	.01333	.00449
.0008	.0006	.0014	.0006	.0016	.0010	.0010	.0008	.00865	.00166
.0012	.0008	.0014	.0006	.0012	.0008	.0008	.0006	.00150	.00083
.0014	.0010	.0050	.0010	.0004	.0002	.0002	.0002	.00333	.00066
.0020	.0014	.0010	.0004	.0002	.0002	.0014	.0004	.00300	.00083
.0022	.0010	.0016	.0002	.0004	.0004	.0004	.0004	.00166	.00066
.0022	.0014	.0052	.0008	.0004	.0002	.0014	.0006	.00133	.00100
.0010	.0008	.0008	.0006	.0003	.0002	.0006	.0006	.00200	.00066
.0020	.0012	.0012	.0004	.0005	.0004	.0014	.0006	.00083	.00033
.0012	.0012	.0030	.0005	.0002	.0002	.0020	.0016	.00133	.00083
.0006	.0004	.0014	.0006	.0006	.0002	.0014	.0006	.00100	.00066
.0020	.0008	.0020	.0008	.0002	.0002	.0012	.0000	.00099	.00166
.0018	.0010	.0020	.0010	.0006	.0004	.0080	.0016	.00250	.00150
.0044	.0028	.0014	.0006	.0002	.0002	.0200	.0054	.00216	.00083
.0014	.0010	.0028	.0012	.0006	.0002	.0020	.0004	.00266	.00083
.0024	.0010	.0010	.0006	.0004	.0002	.0020	.0006	.00216	.00050
.0012	.0012	.0014	.0008	.0006	.0004	.0004	.0004	.00083	.00066
.0052	.0022	.0016	.0002	..	..	.0190	.0044	.00166	.00083
..	..	..	..	..	..	.0100	.0030		
Highest .0052	.0028	.0052	.0012	.0046	.0014	.0200	.0054	.01333	.00449

0.1963 inch square. These do not, however, represent the full quantity of the micro-flaws present in any of the metals observed even within the limited area above mentioned, as measurements were only made on a number of the typical flaws under observation, but there were numerous others present in almost every case. The above observations indicate that micro-flaws to the extent of at least 100 per square inch of surface existed in each metal.

The internal micro-flaws in steel are often so minute as not to materially interfere with the results of the ordinary tensile tests; and as they are not generally easily discernible

SET II.—*Continued.*—TABLE VII.—Measurements of Internal Micro-Flaws in Siemens Steel Steamship Propeller Crankshafts and other Steel Shafts

Siemens Steel Propeller Crankshaft. Index No. 20 A. (See Fig. 20, page 14.)		Soft Siemens Steel Shaft. Index No.21A.* (See Fig. 21, page 15.)		Hard Siemens Steel Shaft. Index No. 22 A. (See Fig. 24, page 16)		Soft Bessemer Steel Shaft. Index No. 23A. (See Fig. 25, page 17.)	
Longitudinal Dimensions of Internal Micro-Flaws.	Transverse Dimensions of Internal Micro-Flaws.	Longitudinal Dimensions of Internal Micro-Flaws.	Transverse Dimensions of Internal Micro-Flaws.	Longitudinal Dimensions of Internal Micro-Flaws.	Transverse Dimensions of Internal Micro-Flaws.	Longitudinal Dimensions of Internal Micro-Flaws.	Transverse Dimensions of Internal Micro-Flaws.
	*Dimensions in Parts of an Inch.*						
.00420	.00083	.0160	.0040	.0060	.0026	.0026	.0010
.00059	.00047	.0020	.0010	.0040	.0010	.0006	.0004
.00170	.00047	.0040	.0014	.0020	.0010	.0006	.0004
.00083	.00059	.0060	.0010	.0120	.0080	.0006	.0004
.00070	.00059	.0014	.0006	.0004	.0002	.0004	.0002
.00083	.00035	.0028	.0018	.0020	.0020	.0003	.0002
.00070	.00059	.0094	.0014	.0028	.0008	.0006	.0004
.00170	.00106	.0058	.0010	.0034	.0012	.0002	.0002
.00210	.00130	.0024	.0002	.0080	.0026	·0020	.0010
.00190	.00083	.0008	.00002	.0040	.0010	.0010	.0008
.00360	.00083	.0030	.0006	.0014	.0006	.0010	.0008
.00140	.00059	.0034	.0006	.0026	.0010	.0006	.0004
.00094	.00083	.0260	.0016	.0024	.0010	.0012	.0004
.00118	.00106	.0054	.0018	.0008	.0006	.0008	.0006
.00083	.00083	.0024	.0004	.0010	.0010	.0010	.0006
.00140	.00083	.0070	.0040	.0030	.0008	.0006	.0004
.00118	.00070	.0070	.0001	.0008	.0004	.0004	.0003
.00170	.00035	.0050	.0016	.0038	.0024	.0002	.0002
.00380	.00047	.0180	.0012	.0064	.0020	.0004	.0002
.00170	.00170	.0120	.0020	.0046	.0006	.0002	.0002
.	..	.0360	.0030				
Highest .00420	.00170	.0360	.0040	.0120	.0080	.0026	.0010

* A micro-flaw of considerable extent was observed in this soft Siemens steel shaft, it being 0.036 in. long and 0.004 in. wide.

without microscopic observation, their presence remains unsuspected unless detected by a careful high-power microscopical examination.

Internal micro-flaws of various character are nevertheless almost invariably present in masses of steel, and constitute sources of initial weakness which not unfrequently produce those mysterious and sudden fractures of steel axles, tyres, and shafts, productive of such calamitous results. A fracture once commencing at one of these micro-flaws (started, probably, by some sudden shock or vibration, or owing to the deterioration by fatigue in the metal), runs straight through

SET II.—*Continued.*—TABLE VIII.—MEASUREMENTS OF INTERNAL MICRO-FLAWS IN VARIOUS STEEL SHAFTS.

Aluminium Steel Shaft. Index No. 24 A.		Nickel Steel Shaft. Index No. 25 A. (See Fig. 26, page 17.)		Copper Steel Shaft. Index No. 26 A. (See Fig. 27, page 18)		Silicon Steel Shaft. Index No. 27 A. (See Fig. 28, page 18.)		Chromium Steel Shaft. Index No. 28 A.	
Longitudinal Dimensions of Internal Micro-Flaws.	Transverse Dimensions of Internal Micro-Flaws.	Longitudinal Dimensions of Internal Micro-Flaws.	Transverse Dimensions of Internal Micro-Flaws.	Longitudinal Dimensions of Internal Micro-Flaws.	Transverse Dimensions of Internal Micro-Flaws.	Longitudinal Dimensions of Internal Micro-Flaws.	Transverse Dimensions of Internal Micro-Flaws.	Longitudinal Dimensions of Internal Micro-Flaws.	Transverse Dimensions of Internal Micro-Flaws.
Dimensions in Parts of an Inch.									
.0006	.0004	.0032	.0010	.0042	.0034	.0056	.0030	.0020	.0010
.0006	.0005	.0008	.0004	.0050	.0030	.0016	.0006	.0006	.0002
.0004	.0002	.0016	.0014	.0014	.0008	.0016	.0010	.0008	.0004
.0006	.0004	.0016	.0010	.0010	.0008	.0026	.0010	.0004	.0003
.0006	.0004	.0020	.0014	.0024	.0016	.0014	.0006	.0006	.0004
.0004	.0004	.0036	.0022	.0016	.0008	.0014	.0006	.0004	.0002
.0012	.0003	.0010	.0014	.0006	.0004	.0025	.0018	.0004	.0002
.0004	.0004	.0006	.0004	.0010	.0014	.0012	.0006	.0006	.0002
.0006	.0002	.0012	.0008	.0024	.0010	.0020	.0010	.0006	.0002
.0006	.0004	.0010	.0008	.0030	.0020	.0014	.0010	.0006	.0002
.0004	.0004	.0004	.0006	.0016	.0008	.0012	.0008	.0008	.0004
.0004	.0002	.0013	.0010	.0036	.0014	.0012	.0010	.0004	.0004
.0002	.0002	.0024	.0014	.0060	.0044	.0010	.0008	.0010	.0006
.0002	.0002	.0026	.0024	.0032	.0020	.0012	.0010	.0006	.0002
.0002	.0002	.0022	.0014	.0024	.0018	.0024	.0016	.0008	.0002
.0004	.0002	.0020	.0008	.0024	.0016	.0010	.0008	.0004	.0006
.0002	.0002	.0010	.0018	.0014	.0008	.0016	.0006	.0006	.0004
.0004	.0004	.0028	.0020	.0022	.0012	.0020	.0012	.0010	.0008
.0004	.0004	.0018	.0012	.0020	.0014	.0014	.0010	.0004	.0004
.0010	.0001	.0018	.0008	.0012	.0010	.0020	.0014	.0008	.0004
		.0016	.0008					.0020	.0010
H guest .0012	.0003	.0040	.0024	.0080	.0044	.0056	.0030	.0020	.0010

a steel forging on the line of least resistance, in a similar manner to the fracture of glass or ice.

Instances of this class of fracture are doubtless afforded by the following sudden breakages of steel axles which have occurred in actual practice, viz.:

1. To the fracture of the driving Siemens steel crank axle of a fast express passenger train, which snapped whilst

running at high speed on a 40-chain curve, this fracture causing the frightful calamity at Bullhouse, near Penistone, on July 16, 1884, attended with serious loss of life. It may be remarked that this axle had not been running many years.

2. To the sudden snapping, into four pieces, of the Bessemer steel leading axle of a fish van, attached to an express

SET III.—TABLE IX.—Measurements of Internal Micro-Flaws in Siemens Steel Ship Plates, Bridge Girder Plates, and Boiler Plates.

Siemens Steel Ship-Plate. Index No. 29 A. (See Fig. 29, page 19.)		Siemens Steel Bridge Girder Plate. Index No. 30 A.		Siemens Steel Boiler Plate. Index No. 31A. (See Fig.30, page 19.)		Siemens Steel Boiler Plate. Index No. 32 A.	
Longitudinal Dimensions of Internal Micro-Flaws.	Transverse Dimensions of Internal Micro-Flaws.	Longitudinal Dimensions of Internal Micro-Flaws.	Transverse Dimensions of Internal Micro-Flaws.	Longitudinal Dimensions of Internal Micro-Flaws.	Transverse Dimensions of Internal Micro-Flaws.	Longitudinal Dimensions of Internal Micro-Flaws.	Transverse Dimensions of Internal Micro-Flaws.
Dimensions in Parts of an Inch.							
.0108	.0008	.0012	.0002	.0018	.0002	.0050	.0002
.0084	.0004	.0008	.0002	.0022	.0001	.0014	.0002
.0020	.0001	.0004	.0001	.0010	.0004	.0032	.0002
.0040	.0001	.0003	.0001	.0046	.0002	.0080	.0002
.0042	.0002	.0008	.0001	.0038	.0002	.0022	.0004
.0026	.0002	.0004	.0004	.0010	.0004	.0030	.0002
.0076	.0002	.0004	.0002	.0014	.0004	.0020	.0004
.0070	.0003	.0006	.0002	.0018	.0004	.0048	.0002
.0064	.0002	.0003	.0002	.0012	.0002	.0016	.0001
.0036	.0001	.0012	.0003	.0012	.0004	.0024	.0002
.0150	.0004	.0006	.0002	.0022	.0002	.0032	.0002
.0060	.0008	.0008	.0002	.0028	.0002	.0026	.0002
.0046	.0001	.0010	.0002	.0060	.0004	.0008	.0004
.0054	.0001	.0016	.0003	.0026	.0002	.0050	.0001
.0040	.0002	.0006	.0002	.0038	.0002	.0012	.0002
.0066	.0002	.0022	.0004	.0036	.0002	.0024	.0001
.0040	.0002	.0006	.0002	.0020	.0002	.0036	.0001
.0060	.0002	.0006	.0002	.0026	.0002	.0026	.0001
.0028	.0001	.0014	.0002	.0030	.0002	.0012	.0001
.0050	.0003	.0006	.0001	.0048	.0001	.0020	.0002
.0024	.0001	.0008	.0002				
Highest .0150	.0008	.0022	.0004	.0060	.0004	.0060	.0004

passenger train, near Winwick Junction, on September 20, 1888, whilst the train was proceeding at full speed over the following curves: "An 80-chain curve to the west, then straight for about 105 yards, then upon a 40-chain curve to the west for about 83 yards, then straight for about 105 yards, then upon a 20-chain curve to the west for about 100 yards, to the trailing points of the junction of the up fast and up slow lines, and then upon a 70-chain curve to the west."

3. To a case of derailment of a passenger train, attended with loss of life and damage to rolling stock, which occurred at Penistone, March 30, 1889. This accident was caused by the instantaneous fracture into several pieces of the straight

TABLE X.—GENERAL SUMMARY OF RESULTS. MEASUREMENTS OF INTERNAL MICRO-FLAWS IN STEEL RAILWAY AXLES STEAMSHIP PROPELLER SHAFTS, &c.

Description.	Index Number.	Longitudinal Dimensions of Internal Micro-Flaws.	Transverse Dimensions of Internal Micro-Flaws.
		Dimensions in Parts of an Inch.	
Bess mer steel railway axle, British make . .	2 A	.0024	.0010
Ditto Ditto . .	3 A	.0300	.0040
Bessemer steel railway axle, Belgian make..	4 A	.0012	.0006
Ditto Ditto . .	5 A	.0036	.0014
Siemens steel railway axle, British make ..	6 A	.0014	.0006
Ditto Ditto . .	7 A	.0100	.0060
Ditto Ditto . .	8 A	.0050	.0014
Ditto Ditto . .	9 A	.0120	.0010
Siemens steel locomotive crank-axle, British make	10 A	.0012	.0006
Bessemer steel railway tyre, British make..	11 A	.0010	.0006
Siemens ,, ,, . .	12 A	.0012	.0008
Bessemer steel railway tyre, Belgian make .	13 A	.0020	.0008
Bessemer steel rail, British make ..	14 A	.0150	.0012
Siemens steel propeller shaft, British make	15 A	.0052	.0028
Ditto Ditto . .	16 A	.0052	.0012
Ditto Ditto . .	17 A	.0046	.0014
Siemens steel warship propeller shaft, British make	18 A	.0200	.0054
Ditto Ditto . .	19 A	.01333	.00449
Siemens steel propeller crankshaft, British make	20 A	.0042	.0017
Soft Siemens steel shaft, British make ..	21 A	.0360	.0010
Hard Siemens steel shaft, British make ..	22 A	.0120	.0080
Soft Bessemer steel shaft, ,, . .	23 A	.0026	.0010
Aluminium steel shaft, ,, . .	24 A	.0012	.0005
Nickel steel shaft, ,, . .	25 A	.0040	.0024
Copper ,, ,, . .	26 A	.0080	.0014
Silicon ,, ,, . .	27 A	.0056	.0030
Chromium steel shaft, ,, . .	28 A	.0020	.0010
Siemens steel ship-plate, ,, . .	29 A	.0150	.0008
,, bridge girder plate, British make ..	30 A	.0022	.0004
Siemens steel boiler plate, British make ..	31 A	.0060	.0004
Ditto Ditto . .	32 A	.0060	.0001

leading cast-steel axle of the engine whilst the train was rounding a curve of about 48 chains radius.

4. Attention may be called to the fracture of the steel crank-axle of a passenger train at Larbert Junction, on the Caledonian and North British Railways, on August 24, 1885. At the place of this accident there is a curve westward of about 20 chains radius.

C

5. Reference may also be made to the fracture of the engine axle of a passenger train from Barnsley to Stairfoot, which broke immediately after leaving the short and somewhat sharp curve at Quarry Junction.

6. To the fracture of the leading tender axle of a goods engine, which broke on September 7, 1887, in the neighbourhood of Wortley, near Sheffield, whilst the train was rounding a sharp curve.

In connection with the accident which occurred at the Saint Vigean's Junction, near Arbroath, on the Dundee and Arbroath Joint Railway, Major Marindin stated, "This is not the only place where curves of such small radius ought to be enlarged. It is impossible to get safety with rigid wheel bases on such curves."

7. Further instances of the liability to sudden fracture of steel railway axles, the danger having probably been accentuated by the alternating torsional strains set up previously by the curves of the road, are afforded by the fracture of the Siemens steel axle of the engine of the London and Scotland passenger train at Kirtle Bridge, near Carlisle, January 8, 1891.

8. The fracture of the Siemens steel axle of the engine of a passenger train, at Abington, Caledonian Railway, February 4, 1891.

9. The appalling accident on the Canadian Pacific Railway, near Schreiber, on February 5, 1891, caused by the fracture of the Siemens steel axle (of British manufacture) of a sleeping car.

10. The fracture of the steel straight driving axle on the engine of the Manchester and Liverpool train, near Warrington, in March, 1891.

11. The fracture of the steel crank axle of the engine of the "Flying Dutchman" express near Reading May 16, 1891.

12. The fracture of the Siemens steel crank axle of the engine of an excursion train, at Tallington, G.N.R., August, 1892.

13. The fracture of a Bessemer steel van axle at Crewe, on January 25, 1894.

14. The fracture of a steel wagon axle, at Wortley, on November 28, 1892.

15. The accident caused by the fracture of the leading straight steel engine axle, at Wood Green, February 28, 1895.

16. The recent sudden fracture of the toughened Bessemer steel passenger carriage axle on the Metropolitan Railway.

17. The sudden fracture of the Bessemer steel rail at St. Neot's, on the Great Northern Railway, November 10, 1895, which wrecked the Scotch express with calamitous results.

18. The fracture of a steel locomotive passenger engine tyre on the London and South-Western Railway, near Chard Junction, December 10, 1895.

19. The fracture of the steel axle of the engine of the express from Carlisle to London, near Brimbo, 30 miles south of Carlisle, February 13, 1896.

20. The fracture recently of a Siemens steel locomotive crank axle, after a life of under $7\frac{1}{2}$ years on one of our main lines.

21. The recent breakage of a Siemens steel locomotive straight engine axle, with a life of only about $1\frac{1}{2}$ years, on a main line of railway.

22. The fact that on the Great Indian Peninsula and Indian Midland Railway Company's, and on other Indian railways, there have been numerous fractured and flawed steel axles.

23. The recent sudden fracture in March last of a Siemens steel locomotive connecting-rod, which broke whilst running at full speed on a London express to the north. The author has made a careful examination into the cause of this breakage. It may be remarked that this steel connecting-rod has only been running about four years.

24. The fracture of a number of Siemens-Martin steel boiler stays on the trial of the boilers of H.M. cruiser Melampus, early in May, 1896. The author made a careful chemical, physical, and microscopical examination into the

cause of this fracture, and furnished a report thereon to the Admiralty.

25. The recent fracture of a steel marine boiler, of which the author has cognisance, the plates of which split across in the ordinary hydraulic testing.

26. The fracture of a steel boiler tube which lately came under the author's notice.

27. The recent fracture of a steamship propeller shaft described by Mr. A. E. Seaton, Institution of Naval Architects, March, 1896.

28. The disaster caused by the bursting of one of the cannon of the French battery ship Bouvines, at Toulon, in July, 1895.

29. The fracture of the propeller shaft of a large steamship in the Atlantic, February 7, 1896.

The author examined several of the above-mentioned fractures by permission (including the steel crank axle which broke at Bullhouse, Penistone, in 1884), or in course of his professional work as a metallurgical testing engineer and expert. He found in these cases that fracture was induced by internal undetectable growing flaws of the nature of those alluded to in this paper, apparently primarily arising from the segregation of impurities in steel.

Additional information of the tendency of steel axles to sudden fracture apparently from the brittleness produced by the deterioration by fatigue, is afforded by the unsatisfactory experience with steel axles on the railways of India (see remarks hereon by Sir Alexander M. Rendel, K.C.I.E., M.A., M. Inst. C.E., *The Engineer*, September 19, 1890); and also on the railways of Australia.

EXAMPLES OF THE SUDDEN FRACTURE OF STEEL PROPELLER SHAFTS, &C.

Reference may also be made in connection with this subject to the following recent sudden and apparently mysterious breakages of steel propeller shafts which further illustrate the danger to steel shafts from torsional strain, viz., the sudden fracture of the propeller shaft of the s.s. Dolphin, United States Navy, which broke on the trial trip. To the sudden fracture of the steel propeller shaft of the s.s. City of Paris, and also to the serious accident which occurred in mid-ocean on December 23, 1892, to the celebrated Cunard steamer the Umbria, caused by the sudden fracture of the steel screw shaft. The recent fracture of a steamship propeller shaft, described by Mr. A. E. Seaton, Institution of Naval Architects, March, 1896. The disaster caused by the bursting of one of the cannon of the French battery ship Bouvines, at Toulon, in July, 1895. The sudden fracture of one of the piston shafts of the torpedo-boat Sarrazin, in 1895. The fracture of the propeller shaft of a large steam-ship in the Atlantic, February 7, 1896. The breaking of the starboard propeller shaft of the Atlantic liner, Paris, on October 8, 1896, which broke in the tube length about 50 ft. from the screw. The ship was about 353 miles from Sandy Hook lightship at the time of the accident. It is reported that the steamer Linlithgow, to Leith from San Francisco, while crossing the Pacific on August 11, 1896, broke her propeller shaft and became helpless.*

Some further idea of the magnitude of the danger arising from the liability of the steel propeller shafts snapping suddenly under torsional strain will be obtained on reference to the information given by Sir Benjamin B. Baker, F.R.S., M. Inst. C.E. (Report of the British Association, 1885, pages 1185 and 1186), who has stated that during the three years, from 1882 to 1885, no less than 228 large steamers were disabled in consequence of broken shafts, and that the longest life of these steel shafts cannot well be put at more than three or four years. The following particulars respecting the fractured steel shaft of the s.s. Dolphin also indicate that a considerable want of homogeneity, consequent on the uneven segregation of sulphur and other elements, is liable to exist in these large steel forgings, which tends to induce initial

---

* The German Atlantic liner, "Furst Bismarck," is reported as having fractured her starboard shaft off Fire Island, on November 6th, 1896.

weakness when they are subjected to external stress. Professor W. C. Unwin, F.R.S., M. Inst. C.E., remarks in his valuable work, " On the Testing of Materials of Construction," page 290, "a steel propeller shaft of the United States despatch-boat Dolphin broke on the trial trip, and test-bars cut from the shaft gave the following results :

—	Elastic Limit.	Tenacity.	Elongation.
	tons per sq. in.	tons per sq. in.	per cent.
From centre of shaft ...	15.2	24.1	2
From surface of shaft ...	14.3	35.7	18

These and other instances prove from actual experience that large steel shafts, however well they may be constructed, are rendered liable to sudden and accidental fracture from the deterioration of the metal consequent on the heavy and repeated torsional and other stresses to which they are perpetually exposed in course of their ordinary wear and tear; the liability to fracture being often increased by the presence of even minute quantities of sulphur and other impurities for reasons previously stated.

Dangerous blow-holes and extensive internal fissures are not infrequently discovered in steel propeller shafts, and also in steel railway axles, both in crank and straight axles, as the experience of many marine engineers and locomotive superintendents shows. A large internal flaw, or hollow cavity, in size about 13½ in. long by 10 in. wide, was recently found in a large steel steamship propeller shaft of 18½ in. diameter (Fig. 31). This flaw was fortunately detected, almost at the last moment, in time to prevent a serious marine calamity (see remarks by Mr. Manuel on "Internal Defects of Steel Propeller Shafts," *Trans. Inst. Marine Engineers*, vol. I., p. 14).

At Penistone, very recently, the journal of a steel axle on a passenger carriage was discovered defective. On examination it was found that the journal of the axle was quite hollow, a large blow-hole or cavity occupying a considerable portion of the interior of the journal. There was only a thin shell of metal holding the journal together, and the defect was fortunately discovered in time, otherwise a serious accident might have occurred. The author recently examined a steel shaft about 6 in. in diameter. About 2 in. from the centre he found a long hollow cavity, or flaw, extending about 5 in. or 6 in., the cavity being nearly ½ in. wide for a distance of 3 in.

Fractures of wrought-iron axles and shafts do not appear, from the author's experience, so liable to occur from the above cause, because any possible internal micro-flaw in the built-up and welded axle or shaft has not a tendency to run across and fracture or destroy the whole forging, but the internal fracture would stop short at the nearest weld junction.

Steel being a cast material, any undetectable flaws, blow-holes, or other imperfections arising from the above-mentioned or other causes in the molten metal, remain therein during the forging, manipulation, and subsequent wear-and-tear of the metal. Further, these structural defects in ingots cannot be welded up or got rid of during the subsequent working or manufacture of steel into axles and shafts, but always remain in the metal as elements of weakness. (See the remarks, accompanied by illustrations on this subject, by Mr. G. W. Manuel, V.P. Inst. Marine Engineers, *Trans. Inst. Marine Engineers*, vol. I., page 14.) The difficulty of obtaining sound homogeneous steel forgings has been referred to by Mr. F. J. R. Carulla, of Derby, at a recent discussion at the Iron and Steel Institute, October, 1891, and in the remarks of Mr. R. M. Deeley in a letter in *The Engineer*, May 22, 1891, on Dr. W. Anderson's paper on " Tests for Steel Used in the Manufacture of Artillery." Mr. C. H. Haswell, of New York, has also made observations showing that steel forgings made " in one piece or mass present the conditions of liability to imperfections in welding, both from a lamination of the structure and *voids in the mass*" (*The Engineer*, April 3, 1891).

Mr. H. A. Ivatt, M. Inst. C.E., recently remarked "the weak point about steel crank axles is their liability to start a crack, and a crack once started is practically impossible to stop " (*The Engineer*, April 3, 1891, page 260).

Other investigators have also noticed this liability to internal defects in steel axles and shafts, the accidental fractures of actual practice generally confirming these observations.

In course of the author's experience, whilst recently testing, on separate occasions, a Bessemer steel axle and a Siemens steel axle, the following unsatisfactory non-homogeneous fractures occurred. Each axle broke into three separate pieces, A, B, and C (see Fig. 32), the piece C weighing in the case of the Bessemer steel axle 10½ lb., and in the case of the

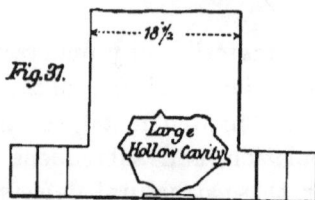

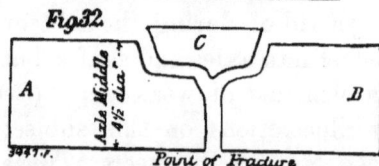

Fig.31.

Large Hollow Cavity

Fig.32.

A   Axle Middle 4½ dia.   C   B

Point of Fracture

Siemens steel axle 6 lb. The piece C in the form shown on Fig. 32 flew off from the axle at the time of the fracture. The fractures of both the Bessemer steel and the Siemens steel axles were similar.

Again, on another recent occasion, whilst a Siemens steel axle was being tested in the usual manner by the drop test, the axle at the fourth blow suddenly snapped in two across the thick boss or wheel seat, at a distance of about 2 ft. 6 in. from the centre or point of impact.

Another Siemens steel axle was taken for testing. The force applied in this instance was the impact from a tup of 1 ton falling from a height of 16 ft. The axle snapped off across the thick boss in a similar manner to the one last mentioned, after having only endured two or three impacts.

In both instances, the circumstance to which attention is drawn is the fact of these axles having broken through the boss, the thickest part of the axle, and far removed from the point of impact. This shows that steel axles are not uniform as regards their resisting strength, and further indicates the non-homogeneous internal physical structure of steel axles. The lack of homogeneity in steel, consequent on the tendency to segregation of the elements in the manufacture of this metal, has been shown by chemical and microscopical tests. This property of steel is productive of internal tension and stress in steel axles and forgings, and assists the destructive action of internal micro-flaws. It has been observed that an analysis of the top of a steel ingot showed combined carbon 0.760 per cent., but the bottom of the same ingot showed only 0.350 per cent. of combined carbon. (Similar uneven segregation was found in the steel rail which fractured at St. Neot's.) The sulphur in the top portion of an ingot was 0.187 per cent., but at the lower part of the ingot it was only 0.044 per cent. The phosphorus in the portion cut from the upper part of the ingot was 0.191 per cent., whereas in the lower part of the ingot the phosphorus was only 0.044 per cent. Sometimes also the outside edge of large steel ingots used for shafts and forgings is found to be very hard, and contains a higher percentage of combined carbon than the interior.

In another instance it has been noticed that six samples taken from the top of a steel ingot, and six samples from the bottom of the same ingot, the samples being taken at equal distances from the outside to the centre of the ingot, showed a diversity in the percentage of combined carbon of about 70 per cent., the percentage of sulphur varying from 0.032 on the outside of the ingot to as much as 0.187 in the centre, the proportion of the phosphorus showing similar variable results. (See also remarks in "Engineering Chemistry," by H. Joshua Phillips, F.R.S.E.)

There is also a tendency in straight steel axles to suddenly fracture simultaneously into several pieces, as exemplified by the fracture into four pieces of the Bessemer steel axle of the fish van in the Winwick Junction accident, September 20, 1888, and again in the case of the broken straight cast-steel engine axle in the Penistone accident, March 30, 1889, which fractured into three pieces. This circumstance affords a further indication of the non-homogeneity of steel axles, and the deterioration of the metal by the fatigue of wear-and-tear, and of the existence of unequal internal stresses in different parts of the axle, inducing fractures in several places at the same time.

Another example of the presence of internal strains in steel forgings will be seen on reference to the illustrated letter thereon by "Marine," in *The Engineer*, October 17, 1890. Indeed, it appears to the author that we are in comparative ignorance of the whole subject of the initial internal stress sometimes obtaining in different parts of the same steel axle or shaft. It is a subject of such importance as to demand careful investigation.

Considerable internal strains have also been found to exist in large masses of steel, such as armour-plates, and these plates have been known to suddenly fracture both before and after fixing on the ship's side, from no apparent causes except those referable to internal initial stress. Examples of this class of fracture were afforded by the failure in this manner of many of the steel armour-plates for the French war vessel, the Terrible. Eighteen out of a total of 90 plates were rejected for cracks resulting from internal strain. Further particulars relating to failure of steel plates from initial stress will be found on reference to *The Engineer*, April 15, 1887, and to the *American Army and Navy Journal*, March 12, 1887.

The author has cognisance of cases of failures of heavy guns apparently traceable to no other cause than initial weakness, consequent on internal strains in the steel (and the presence of minute internal micro-flaws due to sulphur), and Sir Benjamin B. Baker has also alluded to the extent of the deleterious internal strains in gun steel in his remarks just previously quoted.

The author has recently examined the ultimate microscopic structure of two heavy steel gun forgings, near 40 tons in weight. Sulphur areas were found in one of these gun forgings which had been considered defective, the microscope revealing the cause of the defect; the other gun forging was found microscopically to be satisfactory.

The preceding references (indicating the variability of the chemical and physical properties even in the same piece of steel) are made in order to point out from the facts of actual experience, that there is a decided tendency for the components and impurities, such as sulphur, phosphorus, &c., in masses of steel, to segregate in areas, which thus become centres of weakness. This fact is clearly seen on reference to the chemical analyses just quoted, and is still further accentuated by microscopical investigation. Such non-homogeneity in the case of ingot metal tends to accelerate the deterioration by fatigue, and becomes a source of ultimate failure.

The presence of sulphur renders steel short and brittle, owing to the peculiar tendency of this element, when in iron, to segregate as a sulphide of iron along the intercrystalline spaces of the ultimate primary crystals of the metal. This, of course, tends to damage or destroy the tenacity of the steel by reducing the normal or natural cohesion of the constituent primary crystals of the metal. The typical sudden fractures which have recently occurred in steel axles, propeller shafts, rails, &c., may be attributable either to the general unreliability of steel as a material for resisting the shocks arising from the sudden imposition of excessive torsional or other strain incidental to railway axles or shafts, during wear and tear, or to the insidious presence of undetectable voids or other micro-flaws in the metal, similar to those demonstrated in this research, any one of which is capable of becoming a nucleus of destruction in steel. Whatever may be the cause, the fact of the numerous sudden accidental fractures of steel axles and shafts practically demonstrates that there is a treacherous unreliability in this metal for railway axle and shaft service.

In addition to blow-holes, air cavities, treacherous phosphorus and sulphur· areas (reducing the normal metallic cohesion between the facets of the ultimate particles of the metal), and other microscopic flaws, attention should be drawn to another serious and important source of internal weakness in steel, caused by the presence of silicon.

This impurity, during the molten state of the steel, combines with the iron, forming a silicide of iron, which on the cooling of the ingot or mass of steel appears to segregate unevenly in various parts of the metal. Professor Arnold has shown that the normal crystallisation of iron and steel is, moreover, considerably affected by the presence of excess of silicon, the individual primary crystals often exhibiting distorted and somewhat fantastic shapes where silicon is present, and the author's own microscopic experiments have confirmed this. On the cooling of a mass of molten steel, certain portions of this impurity, silicide of iron, locates itself between the facets of the primary crystals of the metal, preventing true metallic contact, and thus reducing the strength of the steel. Other portions also act peculiarly on the iron crystals, or ferrite, portions of mild steels, and mild steel castings, and will sometimes be instrumental in developing, during the cooling of the mass of metal, considerable elongated micro-fissures or cavities along the intercrystalline junctions of the steel. This action, along the intercrystalline spaces of the iron crystals, is sometimes independent of that occurring amongst the carbide areas permeating the mass, but. it is sometimes also associated with it. A further effect of silicon is the frequency with which it is oxidised under pressure in the air cavities or blow-holes, and the latter are consequently often found to be partially filled with a crystalline product from the silicide of iron above referred to.

It is not always easy to detect the difference between internal micro-fissures in steel caused by sulphur and those produced by silicon, but the microscope indicates that the number of micro-flaws in steel due to sulphur greatly exceeds those attributable to silicon. An exceedingly minute percentage of sulphur is capable of developing these internal sulphur flaws in steel.

The elongated fissures produced by the peculiar contractile action of even minute quantities of sulphide and of silicide of iron during the solidification of masses of steel, as just described, are quite distinct from the micro-slag areas occasionally found in large steel forgings, or structures, such as railway axles, tyres, rails, propeller shafts, heavy guns, ship and boiler plates, various steel castings, &c., the fissures attributable to silicide of iron constituting distinct sources of weakness and deterioration, when steel structures are under the fatigue of wear and tear. Professor Arnold has shown that these conditions are especially applicable to steel castings.

A considerable number of the micro-fissures in steel axles, propeller shafts, heavy guns, steel castings, &c., are due to the peculiar action of silicon in various ways, the presence of which in any excess becomes a serious source of danger in steel structures, for reasons briefly explained above. Phosphorus is capable of powerful segregation, but it appears generally more diffusive, and seems frequently scarcely to have such a powerful tendency to segregation as sulphur, though its influence on the intercrystalline junctions of the primary crystals, and on the mass strength of steel and iron, is often decided, when it is present in excess; but, as Professor Arnold has remarked, "sulphur is the more deadly enemy."

Phosphide of iron is less fusible than sulphide of iron, and hence the latter retains its fluid condition longer during the cooling of the mass of steel, and has, consequently, a greater tendency and opportunity to segregation and interdispersion along the intercrystalline junctions of the metal.

The author hopes to be able in the future to communicate the results of further observations on flaws due to sulphur, phosphorus, silicon, and their compounds.

In the course of some experiments by the author on the crystallisation of ice ("Observations on Pure Ice and Snow," Parts I. and II., *Proc. Roy. Soc.*, London, No. 245, 1886, vol. 48, 1890), it was found that not more than about 10 per cent. of the foreign impurities, or saline matters, present in the water or matrix from which the ice crystallised, was taken up by the crystals of the ice on solidification, such impurities as were taken up by the ice crystals being chiefly located between the intercrystalline junctions or spaces. A similar phenomenon occurs in the crystallisation or solidification of masses of liquid steel from high temperature. Hence the reason why the internal micro-flaws, and other sources of initial weakness, resulting from the presence of foreign impurities, are mostly found located in, or adjacent to, the intercrystalline spaces of the ultimate particles of steel. This is the general case, though there are exceptions to this rule.

Although ordinarily these various impurities are not present in sufficient quantity to greatly affect the results of tensile testing (though this is not always the case), yet the existence of these innumerable micro-flaws, or sources of incipient weakness, is a potent factor in the course of a life of wear and tear in promoting the deterioration by fatigue in steel.

The crystalline nature of steel somewhat resembles that of ice, and an internal flaw, however minute, arising from the presence of blow-holes, sulphur areas, cavities, &c., under the influence of prolonged vibration or concussion, is apt, at any moment during the life of steel axles, shafts, or other structures, to extend towards the surface and produce sudden fracture.

It is almost impossible to make steel entirely free from these germs of disintegration, and they being sometimes very minute, are exceedingly difficult of detection. Further, although steel axles, rails, tyres, and propeller shafts, &c. notwithstanding the presence of these ordinarily undetectable flaws, will, in most cases, in the mass, come up to the ordinary tests and requirements demanded by engineers, the author has, in this paper and elsewhere, demonstrated the deleterious presence, form, and general dimensions and peculiarities of many of these insidious micro-germs of metallic disease.

In the examples given, in which the micro-flaws were observed, the forgings met the requirements of the mechanical or other tests imposed, and the metals may be regarded as typical samples of first-class modern steel. It may, therefore, be asked by some that if the ordinary chemical and physical tests are satisfactory, what practical danger is liable to arise from the presence of such minute flaws or defects?

It ought, however, to be observed, that ingot steel is unlike welded wrought iron, and an undetectable internal micro-flaw in steel, however minute, is capable, when steel forgings are under vibratory stress, of becoming a nucleus for the starting of a fracture which would run the whole distance across the diameter of an axle or shaft, as in the case of the fracture of glass or ice. Some recent calamitous accidents, referred to in this paper, have but too strongly accentuated this fact. With fibrous welded wrought-iron forgings the case is different, for reasons previously stated. It was, moreover, noticed in most cases under observation in course of this research, that the normal crystallisation of the metals had to a variable extent been affected in the immediate vicinity of the internal micro-flaws. This is also a result of some importance, and further shows the pernicious influence of these subtle sources of deterioration and fracture in steel axles and shafts, etc.

In addition to the presence of hollow piping holes, blow-holes, gas or air cavities, or other internal flaws of a larger character in steel, this research has demonstrated the existence of numerous internal micro-flaws or germs of metallic disease in various large and important engineering structures of steel, such as steel railway axles, tyres, rails, locomotive crank axles, large propeller shafts, and propeller crankshafts, for the navy and mercantile marine, ship-plates, bridge girder plates, boiler-plates, &c. It will be seen that the investigation has been conducted on an extensive scale, and covers a wide range of materials.

In conclusion, it may be observed that the presence of these numerous micro-flaws in steel is replete with inherent potentialities for evil, tending to facilitate the deterioration by fatigue, to reduce the permanent stability of the forgings, and to promote their premature fracture.

In recording the results of his experimental observations on the microscopical structure of the internal flaws in steel, the author desires it to be understood that, as a metallurgical engineer who has devoted much study to the subject, he yields to no one in appreciation of the admirable qualities and special properties of steel for varied and extensive purposes, though he does not regard it as capable of universal application. It is, moreover, not desirable to be always occupied with contemplating the perfections of a metal, and although the study of imperfections is not a pleasing subject, nevertheless it would be unwise for engineers to ignore the study of the latent faults in steel, or to be devoid of a knowledge of those minute incipient causes which are too often potent in producing the premature deterioration and fracture of steel axles, shafts, and other metallic structures.

There is an old adage that "a man's best friend is he who flatters not, but honestly points out his imperfections;" so also in metallurgy, it is only by carefully studying and recording the imperfections of manipulation and construction that experience is gained which will in the future lead to more perfect results.

For the sake of brevity, Table X., on page 33, gives only the size of one large typical internal micro-flaw in each case. The author, however, measured above 20 micro-flaws of varying size in every metal, within the limited area of about 0.2 square inch, which indicates that micro-flaws to the extent of at least 100 per square inch of surface existed in each metal. This estimate was, however, below the actual number present in each micro-section under observation.

The author has for some time past been conducting a high-power microscopical research on the deterioration by fatigue in various engineering constructions of iron and steel, and hopes ere long to publish the results of his further investigations in this direction.

In addition to the experiments given in the above paper,

the author has recently been engaged on an investigation in connection with the official inquiry of the Board of Trade · into the cause of the fracture of the steel rail whereby the Scotch Express was wrecked at Saint Neot's, on the Great Northern Railway, on November 10, 1895, and some very interesting results have been obtained.

The author has also made physical, chemical, and microscopic examinations of other rails for the Great Northern Railway Company, the Manchester, Sheffield and Lincolnshire Railway Company, and similar examinations of other rails for use in both England and abroad; also of various large propeller shafts, railway axles, tyres, ship-plates, boiler-plates, &c.

He hopes before long to be able to communicate the result of observations on the microscopic structure of locomotive engine axles, steel and iron shafts, chains, &c., which have fractured under known conditions of mileage, age, wear and tear ; also to give the results of a careful study of the microscopic structure of heavy steel ordnance, which latter information may prove interesting to both military and naval engineers.

In connection with this subject, reference may be made to a former paper by the author entitled, " Microscopic Internal Flaws in Steel Rails and Propeller Shafts," (*Engineering*, January 17, 1896, and Messrs. E. and F. N. Spon).

The present research has shown that a careful high-power microscopical investigation on carefully and properly prepared sections, has revealed some of the visible, tangible, and measurable germs of metallic disease influencing the enduring strength of metals. In many instances the author has been able to locate some of the causes of internal weakness in steel and iron shafts, axles, rails, &c., and he hopes ere long to be able to offer suggestions which may assist in reducing the risk of sudden fracture, and in this way to increase the public safety.

APPENDIX A.

MEMORANDA RESPECTING MICRO-SULPHIDE FLAWS AS SEEN
IN MICRO-SECTIONS OF IRON AND STEEL.

The *dark* sulphide of iron flaws are often found in the
ferrite portion of an etched micro-section.

The *dove-coloured* sulphide flaws appear to be manifested
when manganese is present in the steel.

Excess of silicon tends to throw out the sulphide into
meshes, or enveloping membranes.

The presence of manganese tends to throw out the sulphide
as ovoid dove-coloured areas. These in some cases probably
consist of a sulphide of manganese, as the sulphide of iron
flaws are generally of a darker colour.

It seems probable that these micro-sulphide flaws, or
metallic sulphides, in iron and steel, consist of a sub-sulphide
of iron and manganese which forms at a high temperature,
when the combination occurs. Recalescence experiments tend
to prove this.

APPENDIX B.

Reference may be made to the following papers, which
contain information variously connected with this subject:—

"On Tests of Railway Axles," by the Engineers' and
Architects' Society of Austria, published in *The Engineer*,
February 2, 1877, by the late Mr. Walter R. Browne, M.A.

"On Steel for Tyres and Axles," by Sir Benjamin Baker,
K.C.M.G., LL.D., F.R.S.; *Minutes Proc. Inst. C.E.*, vol. lxvii.,
p. 353.

"The Breaking of Locomotive Driving Axles," in *Safe
Railway Working*, by Clement E. Stretton, C.E., p. 96.

. "Observations on the Durability of Wrought-Iron Axles, (see Presidential Address to the Iron and Steel Institute, May, 1889, by Sir James Kitson, Bart. M.P., M. Inst., C.E.); and "Fracture of Steel Axles on Rounding Curves," by Thomas Andrews, in *The Engineer*, June 20, 1890.

"Effect of Temperature on the Strength of Railway Axles," Parts I., II., III.; *Minutes Proc. Inst. C.E.*, vol. lxxxvii., Session 1886-87, vol. xciv., Session 1887-88, and vol. cv., Session 1890-91.

* "Tensile Strength of Wrought-Iron Railway Axles," by Thomas Andrews, in *The Engineer*, December 29, 1893.

* "Effect of Strain on Railway Axles and a Minimum Flexion Resistance Point in Railway Axles," by Thomas Andrews ; *Transactions of the Society of Engineers*, London 1895.

* "Microscopic Observations on the Deterioration by Fatigue in Iron and Steel," by Thomas Andrews. (*Engineering*, 1896.)

* "Microscopic Internal Flaws in Steel Rails and Propeller Shafts," by Thomas Andrews ; *Engineering*, January 17, 1896, and E. & F. N. Spon, London.

* "Iron or Steel Railway Axles ; a Defence of Wrought-Iron Axles and Shafts," by Thomas Andrews.

* "Remarks on Steel Railway Axles," by Thomas Andrews, 1895.

"The Micrographic Analysis of Metals," by Professor J. O. Arnold ; *Iron and Coal Trades Review*, January 31, 1896 ; and the "Influence of Small Quantities of Impurity on Gold and Copper," by Professor J. O. Arnold ; *Engineering*, February 7, 1896.

* "The Micro-Physical Structure of Heavy Steel Guns," by Thomas Andrews.

* "Microscopic Structure, Chemical Specification, and Physical Tests of Steel Rails," by Thomas Andrews.

* "Safe Specification for Steel Locomotive Connecting Rods and Forgings," by Thomas Andrews.

* "Relative Effect of Hydraulic and Impact Forging on Steel Shafts," by Thomas Andrews.

* "Effect of Chilling on the Impact Resistance of Metals," by Thomas Andrews; *Minutes Proc. Inst. C.E.*, vol. ciii., page 231.

* "The Life of Railway Axles," by Thomas Andrews, in *The Engineer*, February 22, 1895.

* "Remarks on the Influence of Temperature on Steel Railway Axles," by Thomas Andrews; *Engineering*, July 5, 1895.

"Micro-Metallography of Iron," by Thomas Andrews; *Proc. Royal Society*, London, 1895.

* A copy of these works may be obtained on application to Thomas Andrews, Wortley, near Sheffield.

LONDON:
PRINTED AT THE BEDFORD PRESS, 20 AND 21, BEDFORDBURY, W.C.

www.ingramcontent.com/pod-product-compliance
Lightning Source LLC
Chambersburg PA
CBHW022017190326
41519CB00010B/1550